Collins

Mental Maths

Ages 5–6

Concepts tested

Concepts	Tests
Counting	1–3, 5, 6, 10, 11, 13–29, 36, 39
Counting in twos	9, 18, 24, 25, 27, 30, 31, 33, 38
Counting in tens	24
Counting in fives	23, 24, 28, 29, 40
Counting in fours	33
More / less	1–6, 8, 9, 14, 17–22, 24, 29, 32–35, 39
Number lines	2, 9–24, 26–29, 33–35, 38
Number sequences	10–13, 16, 18, 21, 23–25, 27–29, 31, 33, 38, 40
Addition	All tests
Subtraction	10–12, 14, 16–18, 22–28, 36, 39, 40
Zero	1, 3, 20, 21, 29
Groups / sets	14, 26, 33
Place value	4, 6, 9, 10, 15, 19, 22, 25, 31–38, 40
Fractions	5, 7, 13, 16, 19, 23, 26–29, 34, 35, 37, 40
Ordinal numbers	7–10
> and <	10–15, 23, 26
Heavier / lighter	2, 28, 29, 31, 37, 38, 40
Longer / shorter / taller	6, 9, 10, 15, 20–22, 31–39
Arbitrary units: length	3, 4, 12, 36
Before / after	6–8
Time: o'clock	2–19, 27, 37–40
Time: half past	18, 23, 28–34
Time: quarter past	30, 35, 36
Time: quarter to	37, 38, 40
Calendar	32, 38, 39
Plane shapes	1–6, 11, 12, 15, 19, 20, 22, 24, 26, 28, 30, 35–37, 39
Solid shapes	29, 35
Graphs	8, 30–34

Introduction

About this book

- This book is part of *Collins Mental Maths*, a series to support your child's development of primary mathematical skills at home.

- The grid on page 2 shows which concepts are tested in this book.

- The 40 tests provide a fun and instant way of testing your child's understanding of these concepts on a weekly basis.

- The questions are presented in a variety of simple styles to make them accessible and engaging, even to more reluctant learners.

- The tests become progressively more challenging, supporting steady advancement.

- To ensure some success every time, each test has two questions at a lower level than the rest.

- Your child may need some help with the early tests in order to get used to the format, but the consistent way in which they are presented should ensure he or she soon becomes familiar with them.

- Answers to all the questions in this book can be found on pages 45–48. There is one mark available for each question.

Recording progress

- At the end of each test is a space for you or your child to record the number of correct answers.

- 'How well did I do?' on page 4 consists of a chart on which your child can record their test marks. This not only involves your child in monitoring his or her progress but also provides practice in handling data for a purpose and involves them in the use of mathematics in an everyday situation. It also provides you with a simple visual record of how your child is meeting the demands of the tests.

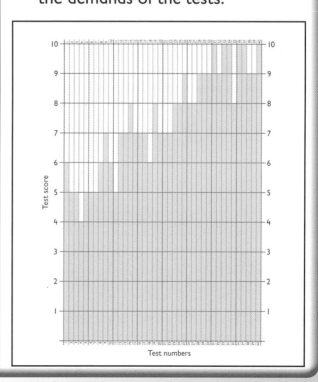

How well did I do?

Shade in your test scores on this chart.

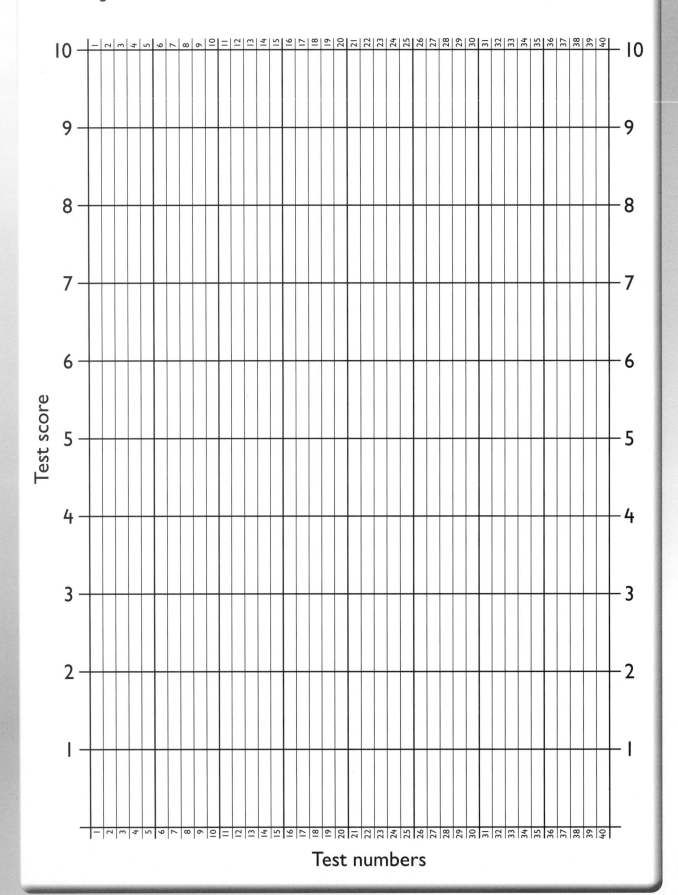

Test numbers

Test 1

1 Join the dots.

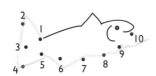

2 I more than 4 is 5.

I more than 7 is

3 How many apples?

4 How many ducks?

5 How many circles?

6

6 Who has more?

7 Who has more?

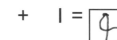

8

● ● ● ● ● 　 ●

4 + 1 = 5

● ● ● ● ● ● ● 　 ●

7 + 1 = 8

9

3 + 2 = 5

10 Join the dots.

Circle

Score

5

Test 2

1 Shade the circles.

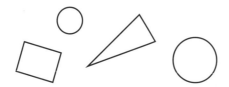

2 1 more than 3 is 4.

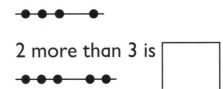

2 more than 3 is ☐

3 Fill in the missing number.

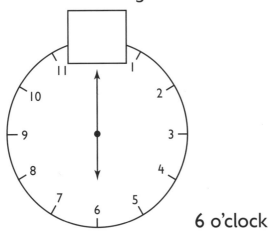

6 o'clock

4 Draw 2:00

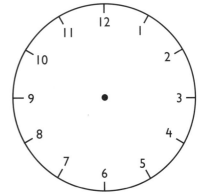

5 Finish the number line.

1	2	3								

6 Draw a ring around the lighter boy.

7 Fill in the number.

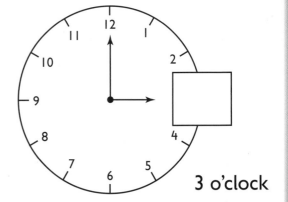

3 o'clock

8

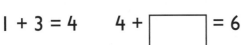

1 + 3 = 4   4 + ☐ = 6

9 2 more than 4 is ☐

10 3 + 2 = ☐

Score

6

Test 3

1  ☐ o'clock

2 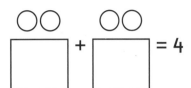 ☐ + ☐ = 4

3  3 fingers long.

Use your own finger to measure.

☐ fingers long.

4 How many circles?

 ☐

5 **4:00** is 4 o'clock.   is 11 o'clock.

6

☐ + ☐ = 7

7  ☐ o'clock

8 2 more than 5 is ☐

1	2	3	4	5	6	7	8	9	10

9

4 + 0 = ☐

10 4 + 2 = ☐

Score

Test 4

1 2 more than 7 is ☐

2 6 + ☐ = 8

3 ○○○   ○○○

　　3　　+　　☐ = 6

4

☐ o'clock

5 2 more than 6 is ☐

| 1 | 2 | 3 | 4 | 5 | 6 | 7 | 8 | 9 | 10 |

6 Join the dots.

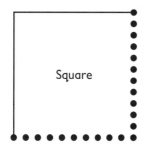

Square

7 How many 10s and units?

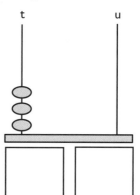

t　　　　u

8 Use your finger to measure.

☐ fingers long.

9 6 + ☐ = 6

10 7 + ☐ = 9

Score

Test 5

1 8 + ☐ = 10

2 6 + ☐ = 10

3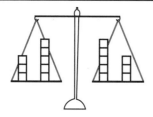

3 + ☐ = 5 + 3

4 Shade the squares.

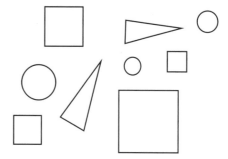

5 2 more than 8 is ☐

| 1 | 2 | 3 | 4 | 5 | 6 | 7 | 8 | 9 | 10 |

6 Draw a ring around the set with $\frac{1}{2}$ shaded.

7

Pat   Jim

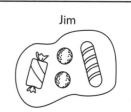

☐ has less.

☐ has more.

8 10 o'clock is ☐ **:**

9 How many squares? ☐

10 4 + 3 = ☐

Score

9

Test 6

1 Who has less?

has less. [box]

2 4 + 4 = [box]

| 1 | 2 | 3 | 4 | 5 | 6 | 7 | 8 | 9 | 10 |

3

3 o'clock [box] o'clock

4 Shade the longest belt.

5 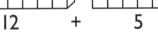 12 + 5

   t   u
= [box] [box]

6 How many triangles?

[box]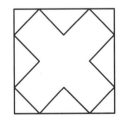

7  [box] o'clock

The car is before the truck.

van   bus   car   truck   bike

8 The van is before the [box]

9 The bus is before the [box]

10 The truck is before the [box]

Score

10

Test 7

1 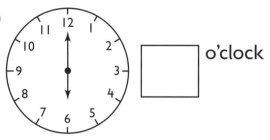 ☐ o'clock

2 6 + 1 = 1 + ☐

3   5 + 5 = ☐

4 Draw a ring around $\frac{1}{2}$ of this set.

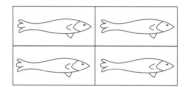

The bus is after the van.

5 The car is after the ☐

van   bus   car   truck   bike

6 The truck is after the ☐

7 The bike is after the ☐

The 1st is before the 2nd.

8 The 2nd is before the ☐ rd.

9 The 5th is before the ☐ th.

10 The 9th is before the ☐ th.

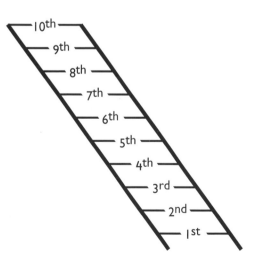

Score

Test 8

1 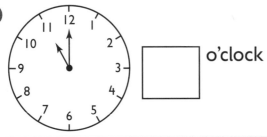 ☐ o'clock

2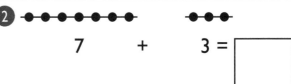

7 + 3 = ☐

The 2nd is after the 1st.

3 The 5th is after the ☐ th.

4 The 8th is after the ☐ th.

5 The 10th is after the ☐ th.

10th
9th
8th
7th
6th
5th
4th
3rd
2nd
1st

Geeta Tom Jill

6 Who has the most?

☐

7 Who has the least?

☐

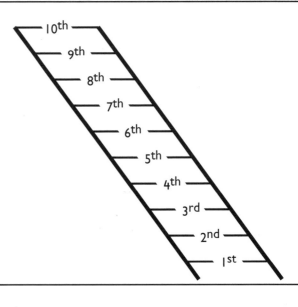

Jack   Tom   Amy   James

FINISH

Jack is 1st.

8 ☐ is 2nd.

9 ☐ is 3rd.

10 ☐ is 4th.

Score

Test 9

1 Fill in the missing number.

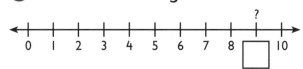

2

t   u

10 + 4 = ☐ ☐

3

One less than four is three.
One less than seven
is ☐

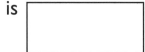

4 Tom is the tallest.

☐ is the tallest.

Grace is 1st in line.

5 John is ☐ in line.

6 Grant is ☐ th in line.

7 May is ☐ th in line.

8 Two less than six is ☐

9 Shade the 3rd horse.

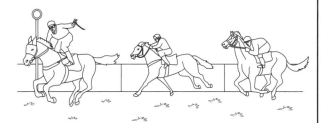

10

☐ o'clock

Score

13

Test 10

1 6 + 3 = ☐

2

☐ > ☐

3

Tim   Jane   Ann

☐ is the tallest.

4

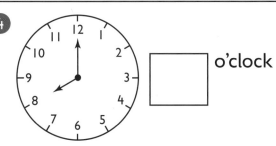

☐ o'clock

5

Shazia   Shane   Helen

FINISH

☐ is 2nd.

6

20   +   9   =

t   u

☐   ☐

7

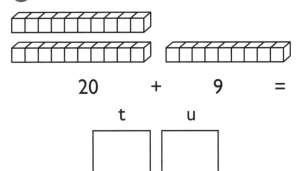

3 − 2 = 1

4 − 1 = ☐

8 Fill in the missing numbers:

1, 2, 3, 4, ☐ , ☐ , 7, 8

9 7 + 3 = ☐

10

3 > ☐

Score

14

Test 11

1

$$2 \quad > \quad \boxed{}$$

2 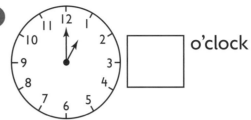 $\boxed{}$ o'clock

3 How many squares? $\boxed{}$

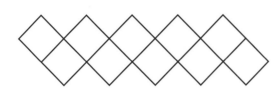

4 Fill in the missing numbers.

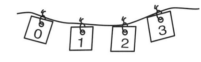

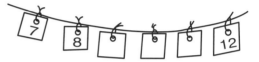

5 $6 + 4 = \boxed{}$

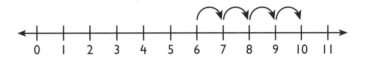

6 Take two from six.

$$6 - 2 = \boxed{}$$

7

$$5 - 1 = \boxed{}$$

8 Fill in the missing numbers.

9 $4 + 6 = \boxed{}$

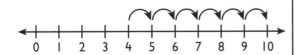

10

$$4 \quad > \quad \boxed{}$$

Test 12

1 Fill in the missing numbers.

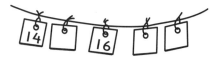

2

 o'clock

3 Take 4 from 6.

$6 - 4 = $ ☐

4 Use your finger to measure.

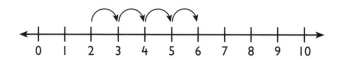

☐ fingers long.

5 $2 + $ ☐ $= 6$

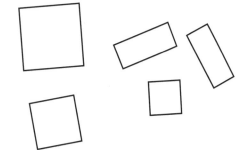

6 Shade the squares.

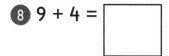

7 Draw a ring around yes or no.

4 is more than 3. (Yes)   No

5 > 6   Yes   No

9 > 10   Yes   No

8 $9 + 4 = $ ☐

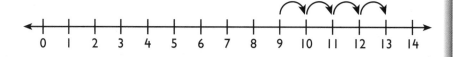

9 $4 + $ ☐ $= 7$

10 Take 3 from 7.

$7 - 3 = $ ☐

Score

Test 13

1 Fill in the missing numbers.

2 $3 +$ ☐ $= 6$

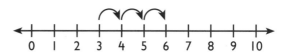

3 Is this correct? Draw a ring around yes or no.

$3 > 8$

Yes   No

4 Draw a ring around $\frac{1}{2}$ of the cars.

5

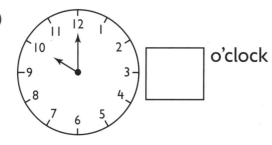

11   Draw a loop around 12.

6 ☐ o'clock

7 $5 +$ ☐ $= 8$

8 Draw a loop around 13.

9 Is this correct? Draw a ring around yes or no.

$12 > 11$

Yes   No

10 $9 +$ ☐ $= 14$

Score

Test 14

1 Draw 3 o'clock.

2 Fill in the number.

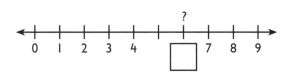

3 4 less than 9 is

4 Is this correct? Draw a ring around yes or no.

$$12 > 13$$

Yes   No

5 Draw a ring around 15.

6 Draw nine o'clock.

7 $5 +$ [ ] $= 9$

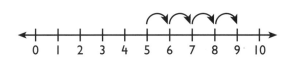

8 Draw a loop around 16.

9 This is a pair. How many shoes are in ten pairs?

10 Take 3 from 8.

$8 - 3 =$

Score

18

Test 15

1 5 + ☐ = 7

2 Draw 6 o'clock.

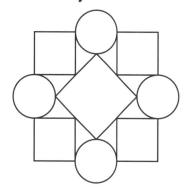

3 Is this correct? Draw a ring around yes or no.

16 > 14

Yes   No

4 Shade the squares.

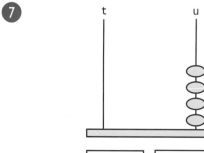

5 Draw a loop around 18.

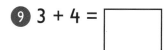

6 Shade the longest bar.

7

8 Draw a loop around 19.

9 3 + 4 = ☐

10 1 + ☐ = 7

Score

Test 16

1 Draw 4 o'clock.

2 Fill in the number.

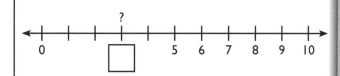

3 How many cubes?

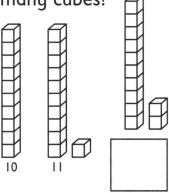

4 Draw 7 o'clock.

5 Take five from ten.

$10 - 5 =$ ☐

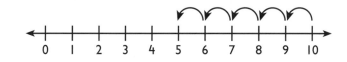

6 Draw a ring around $\frac{1}{2}$ of the boats.

7

12 = 10 + 2

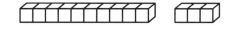

13 = 10 + ☐

8 Fill in the missing numbers.

12, 13, ☐ , ☐ , 16

9 $4 + 6 =$ ☐

10 $5 +$ ☐ $= 10$

Score

Test 17

1 Who has more than Roop?

 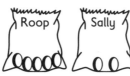

2 Who has less than Roop?

3 Fill in the number.

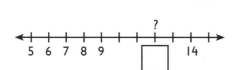

4 How many cubes?

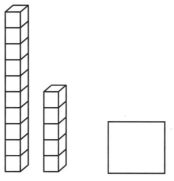

5 3 + 7 =

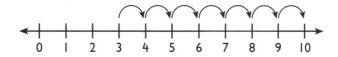

6 Draw 1 o'clock.

7 10 − 7 =

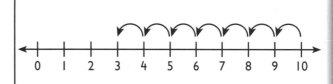

8

15　　　=　　　10　　　+

9 Who has the most?

10 Who has the least?

Test 18

1 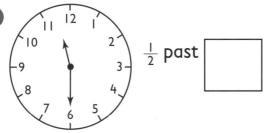 $\frac{1}{2}$ past ☐

2 $10 - 4 =$ ☐

3 How many cubes?

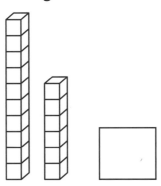

4 Draw 5 o'clock.

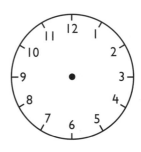

5 Fill in the missing numbers.

2, 4, ☐, ☐, 10, 12

6 $6 +$ ☐ $= 10$

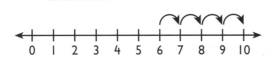

7 Draw 8 o'clock.

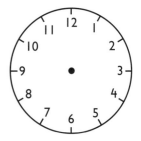

8

16 = 10 + ☐

9 Who has the least? ☐

10 Who has the most? ☐

Test 19

1 Join the dots.

Triangle

2 Draw 2 o'clock.

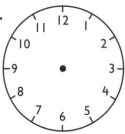

3 How many cubes?

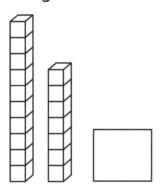

4

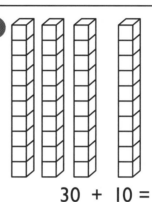

$30 + 10 =$

t   u

5 3 less than 10 is

0 1 2 3 4 5 6 7 8 9 10

6 Shade the triangles.

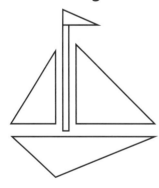

7 Draw a ring around the set with $\frac{1}{2}$ coloured.

8

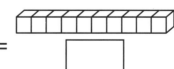

17

=

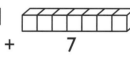

+   7

9 Who has the most?

10 Who has the least?

 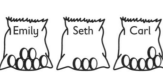 Emily Seth Carl

Score

Test 20

1

$$3 + 0 = \boxed{}$$

2 $0 + \boxed{} = 4$

3 How many cubes?

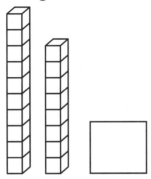

4 Shade the triangles.

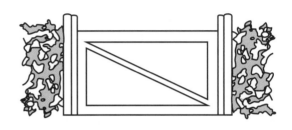

5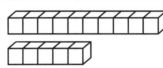

$$10 + 15 = \begin{array}{cc} t & u \\ \boxed{} & \boxed{} \end{array}$$

6 Fill in the number.

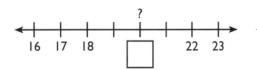

16 17 18 $\boxed{}$ 22 23

7 Shade the bars that are the same length.

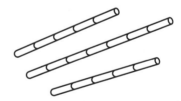

8 18 $= \boxed{} + 8$

9 Who has the least?

$\boxed{}$

10 Who has the most?

$\boxed{}$

 Amaya Graham Peter

Test 21

1 5 + 0 = ⬜

2 1 more than 12 is ⬜

8 9 10 11 12 13 14

3 How many cubes?

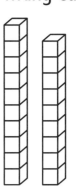

⬜

4 1 less than 12 is ⬜

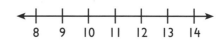
8 9 10 11 12 13 14

5 7 + 6 = ⬜

0 1 2 3 4 5 6 7 8 9 10 11 12 13 14 15

6 Draw a ring around the tallest tree.

7 Fill in the missing numbers. Count in 2s.

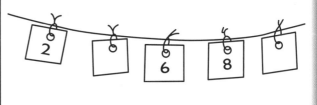
2   ⬜   6   8   ⬜

8

19   =   ⬜   +   9

9 Who has the most?

⬜

10 Who has the least?

⬜

 Jane Anil Ann

Score

25

Test 22

1 2 more than 10 is ☐

2 2 less than 10 is ☐

7 8 9 10 11 12

---

3 How many cubes?

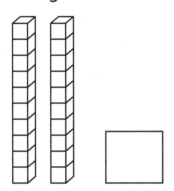

☐

4 Shade the triangles.

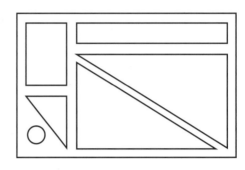

5

18   +   10   =   ☐ t ☐ u

---

6 Who has two less than Dennis?

☐

 Dennis   Deepak   Paula

7 Fill in the number.

?
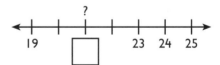
19   ☐   23 24 25

8
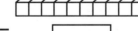
20   =   ☐   +   ☐

---

9 Shade the shortest bar.

10 20 − 3 = ☐

Test 23

1 Fill in the missing numbers. Count in 5s.

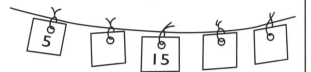

5   15

2 Is this correct? Draw a ring around yes or no.

21 > 19

Yes   No

3

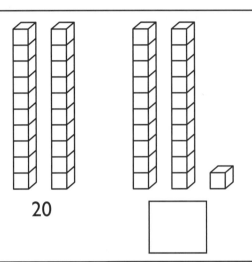

20

4 Take 2 from 12.

12 – 2 =

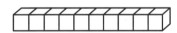

5 Fill in the number.

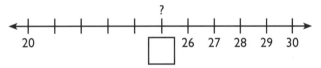

20   ?   26   27   28   29   30

6 Draw a ring around ½ of the balls.

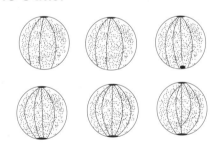

7 How many cubes?

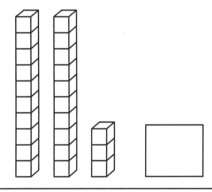

8 Is this correct? Draw a ring around yes or no.

11 > 13   Yes   No

9

   past

10 12 + 4 =

Score

Test 24

1 Fill in the missing numbers.

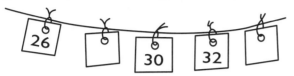

2 Count on in 10s.

10, 20, ☐ , ☐ , 50

3 How many cubes?

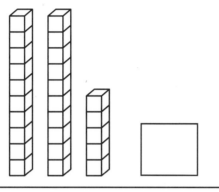

4 Take 3 from 13.

13 − 3 = ☐

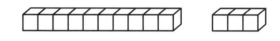

5 10 + 5 = ☐

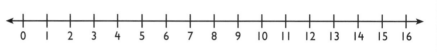

6 Draw a ring around the one that holds the most.

7 How many triangles?

8 Count on in 5s.

20, 25, ☐ , ☐ , 40

9 Take 5 from 15.

15 − 5 = ☐

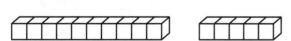

10

1 + 2 + 3 = ☐

Score

28

Test 25

1 10 − 9 = ☐

2 Fill in the missing numbers.

11, ☐, 13, ☐, ☐

3 Take 8 from 18.

18 − 8 = ☐

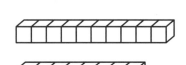

4 How many cubes?

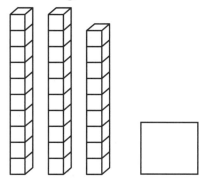

☐

5

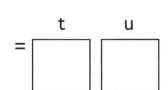

5   +   15   =

t	u
☐	☐

6

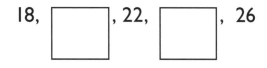

19   +   1 =

t	u
☐	☐

7 Fill in the missing numbers. Count in 2s.

18, ☐, 22, ☐, 26

8

3   +   2   +   1   =   ☐

9

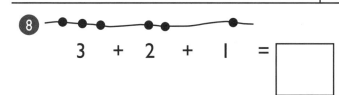

10   +   ☐   = 18

10 10 + ☐ = 19

Score

Test 26

1 Shade the squares.

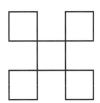

2 How many triangles?

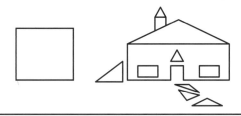

3 Draw a ring around $\frac{1}{2}$ of the fish.

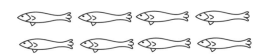

4 Take 10 from 11.

$11 - 10 =$ ☐

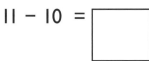

5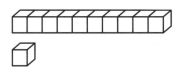

Four stools have ☐ legs.

6 How many cubes?

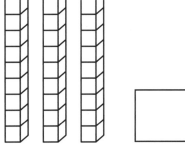

7 How many cubes?

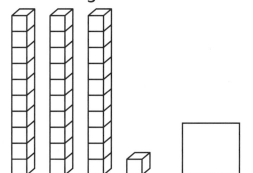

8 $2 + 3 + 4 =$ ☐

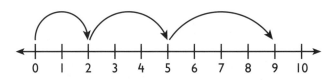

9 Take 10 from 12.

$12 - 10 =$ ☐

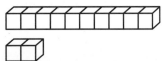

10

☐ < 2

Score

Test 27

1 Fill in the number.

30 31 32 33 [ ] 39 40

2 Take 10 from 17.

17 − 10 = [ ]

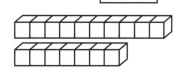

3 Draw a ring around $\frac{1}{2}$ of these.

4 How many cubes?

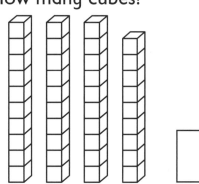

5 Draw a ring around the orange cut in $\frac{1}{2}$.

6

[ ] o'clock

7 How many cubes?

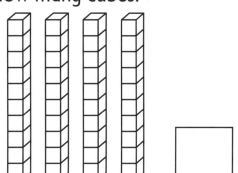

8 Count on in 2s.   30, 32, [ ] , [ ] , [ ]

9 8 − 6 = [ ]

10 Take 10 from 16.

16 − 10 = [ ]

Score

Test 28

1 Draw a ring around the heaviest thing.

2 $4 + 3 + 2 =$ ☐

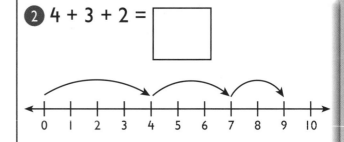

3 Fill in the missing numbers.

5, 10, 15, 20, 25, 30, ☐ , ☐ , ☐ , ☐

4 Show | **5:30** |

5 $3 + 5 =$ ☐

$5 + 2 =$ ☐

6 Shade the triangles.

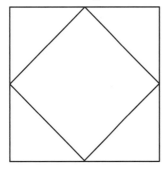

7 How many cubes?

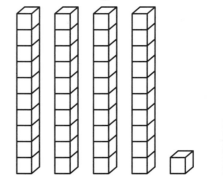

8 $3 + 3 + 3 =$ ☐

9 Draw a ring around the apple cut in $\frac{1}{2}$.

10 $19 - 10 =$ ☐

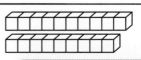

Score

Test 29

1 Fill in the number.

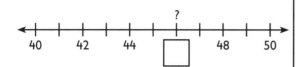

2 Shade the cylinder.

3 Draw a ring around the square cut in $\frac{1}{2}$.

4

$\frac{1}{2}$ past ☐

5

☐ past ☐

6 Which is heavier, A or B?

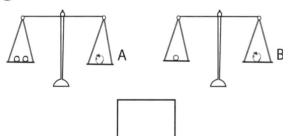

☐

7 10 more than 20 is ☐

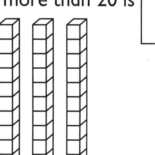

8 Count on in 5s.   25, 30, 35, ☐ , ☐

9

Shade $\frac{1}{2}$

10 0 + 5 = ☐

0 + 2 = ☐

Score

33

Test 30

1  [ ] past [ ]

2 How many triangles?
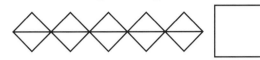 [ ]

3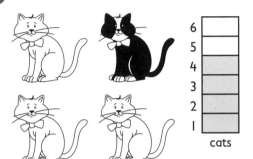
cats

Shade a box for each dog.
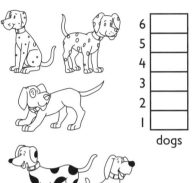
dogs

4 4 + 6 = [ ]   6 + 4 = [ ]

5 3 + 4 + 3 = [ ]

6
[ ] past [ ]

7 How many cubes?
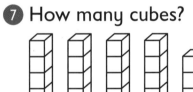 [ ]

8 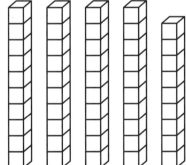 6 legs   [ ] legs

9 $\frac{1}{4}$ past [ ]

10 How many triangles?
 [ ]

Score

Test 31

1 3 + 7 = [ ]   7 + 3 = [ ]

2  $\frac{1}{2}$ past 10 is

[:]

3 Which is heavier, A or B?

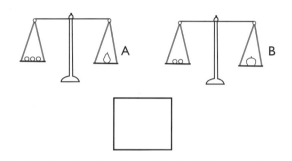

[ ]

4 Shade a box for each fruit.

apples

 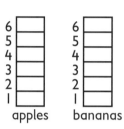

bananas

5 Count on in 2s.   2, 4, [ ], [ ], 10

6

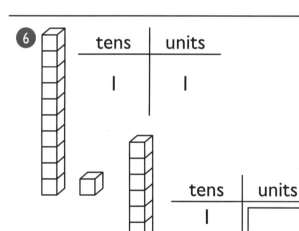

tens	units
1	1

tens	units
1	[ ]

7

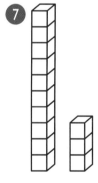

tens	units
[ ]	[ ]

8 Shade the shorter snake.

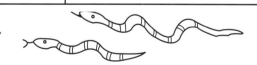

9  [ ] past [ ]

10 4 + 4 + 2 = [ ]

Score

35

Test 32

1 Shade the longer snake.

2 $4 + 5 = \boxed{}$ $5 + 4 = \boxed{}$

3

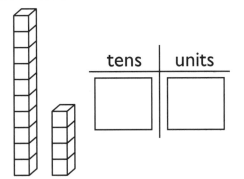

tens	units

4 10 less than 30.

$30 - 10 = \boxed{}$

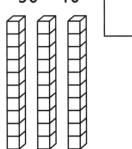

5 How many days are in 3 weeks? $\boxed{}$ days

6

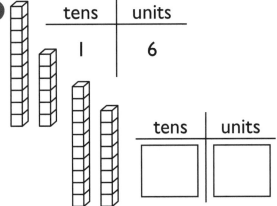

tens	units
1	6

tens	units

7

$\boxed{}$ past $\boxed{}$

8 Who has the least? $\boxed{}$

Sue   Andy   Govind

9 How many sweets have Sue, Andy and Govind altogether? $\boxed{}$

10 Who has more, Sue or Govind? $\boxed{}$

Test 33

1 Shade the longer car.

2

Five birds have ☐ legs.

3 Who has the most?

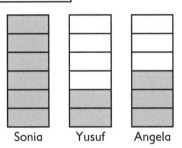

Sonia   Yusuf   Angela

4

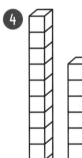 tens | units

☐ | ☐

5 Count on in 4s.   0,   4,   ☐ ,   ☐ ,   16

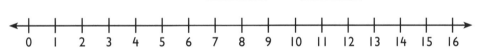

6

☐ past ☐

7

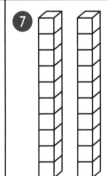 tens | units

☐ | ☐

8 Shade the shorter pencil.

9 Ten more than 95 is ☐

10 7 + 2 = ☐

Score

37

Test 34

1 Shade the longer bar.

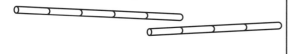

2 What number is 1 less than 100?

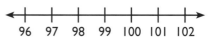

3 Draw $\frac{1}{2}$ past 4.

4 How many has Zahida?

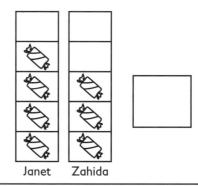

Janet Zahida

5 Look at question 4.
How many more has Janet than Zahida?

6

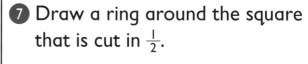

tens | units

7 Draw a ring around the square that is cut in $\frac{1}{2}$.

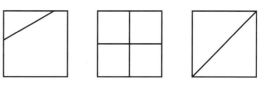

8 Two bags have a total of ☐ marbles.

9 Draw $\frac{1}{2}$ past 8.

10 7 + 1 = ☐

Score

38

1 Shade $\frac{1}{2}$.

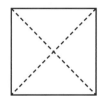

2 3 less than 200 is ☐

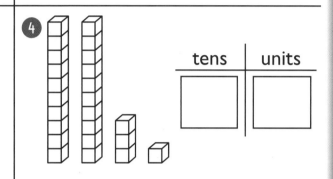

196 197 198 199 200 201

3 Draw $\frac{1}{4}$ past 9.

4

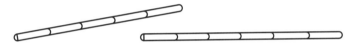

tens	units
☐	☐

5 Shade the shorter bar.

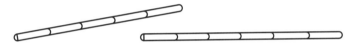

6 Rosie is taller.

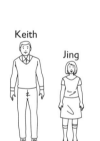 

Keith   Jing   Rosie   Paul

☐ is taller.

7 Which is a cylinder?

(a)    (b) 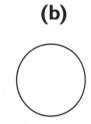   (c)

☐

8 3 + 6 = ☐   6 + 3 = ☐

9 Join the dots and shade.

Rectangle

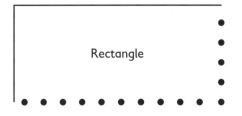

10

$$26 = $$

tens	units
☐	☐

Score

Test 36

1 3 + 5 = ☐

2 Shade the taller person.

3 How many rectangles?

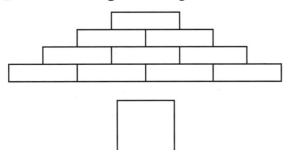

☐

4

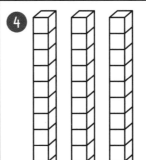

tens	units
☐	☐

5 Take 100 from 400. ☐

6

 is `11:15`

 is ☐ **:**

7 Use your finger to measure.

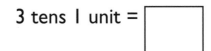

☐ fingers long.

8

t	u
3	0

3 tens 0 units = 30

3 tens 1 unit = ☐

9 Draw ¼ past 10.

10 6 − 1 = ☐

Score

Test 37

1 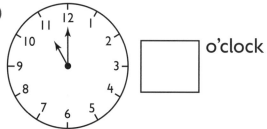 ▢ o'clock

2 Shade the rectangles.

3
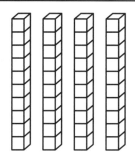

tens	units

4

Charlotte is heavier.

▭ is heavier.

5 Draw a ring around the rectangle cut in $\frac{1}{2}$.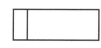

6 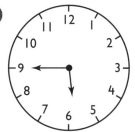 $\frac{1}{4}$ to ▢

7  $\frac{1}{4}$ to ▢

8 ▭ is taller.
 Ross Lauren

9 4 tens 9 units = ▭

10 25 + 2 = ▢

Score

Test 38

1 Shade the longer bar.

2 Show $\frac{1}{4}$ to 7.

3

3 o'clock

3:00

4 o'clock

:00

4 Draw a ring around the heavier apple.

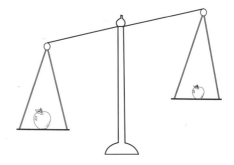

5 **8:00** 8 o'clock   **12:00** [ ] o'clock

6

1 + 3 = 3 + [ ]

7 Count back in 2s.

20, 18, [ ] , [ ] , 12, 10

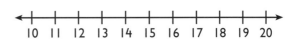

8 There are [ ] days in a week. (3, 7 or 10)

9 28 = [ ] tens [ ] units

10 4 + 4 =

[ ]

Score

Test 39

1 Shade the shortest bar.

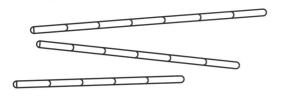

2

:00

3

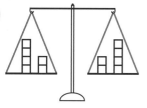

4 + 2 = 2 + ⬜

4 Shade the rectangles.

5 There are ⬜ months in 1 year. (7, 12 or 24)

6 10 more than 30 is ⬜

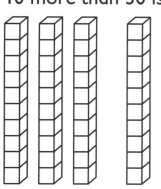

7 3 + 4 = ⬜ + 3

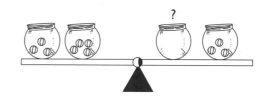

8 Draw a ring around the taller tree.

9

:00

10 10 − 5 =

⬜

Score

Test 40

1 10    13   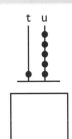

2 Shade $\frac{1}{2}$.

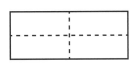

3 Draw a ring around the heavier animal.

4
| 10:00 | is | |

| 5:00 | is | 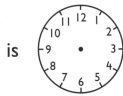 |

5 Fill in the missing numbers.

45, 50, 55, ☐, ☐, ☐, 75

6 quarter to 3

 quarter to ☐

7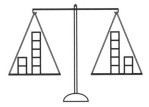

2 + 5 = 5 + ☐

8 How many minutes are in 2 hours? ☐

9 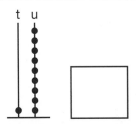 ☐

10 4 − 1 =

☐

Score

Answers

Test 1
1.
2. 8
3. 3; 7
4. 0; 2
5. 6
6. Pat
7. Jill
8. 8
9. 5
10. Circle

Test 2
1.
2. 5
3. 12
4.
5. 4, 5, 6, 7, 8, 9, 10
6.
7. 3
8. 2
9. 6
10. 5

Test 3
1. 4
2. 2 + 2
3. Approx. 4
4. 10
5. 11:00
6. 5 + 2
7. 3
8. 7
9. 4
10. 6

Test 4
1. 9
2. 2
3. 3
4. 9
5. 8
6. Square
7. t: 3; u: 0
8. Approx. 3
9. 0
10. 2

Test 5
1. 2
2. 4
3. 5
4.
5. 10
6.
7. Pat has less; Jim has more
8. 10:00
9. 8
10. 7

Test 6
1. Sitar
2. 8
3. 8
4.
5. t: 1; u: 7
6. 8
7. 3
8. bus
9. car
10. bike

Test 7
1. 6
2. 6
3. 10
4. A ring should be drawn around any two fish.
5. bus
6. car
7. truck
8. 3
9. 6
10. 10

Test 8
1. 11
2. 10
3. 4
4. 7
5. 9
6. Jill
7. Tom
8. Tom
9. Amy
10. James

Test 9
1. 9
2. t: 1; u: 4
3. six
4. Sue
5. 2nd
6. 5
7. 4
8. four
9.
10. 5

Test 10
1. 9
2. 6 > 5
3. Ann
4. 8
5. Shane
6. t: 2; u: 9
7. 3
8. 5, 6
9. 10
10. 2

Answers

Test 11
1. 1
2. 1
3. 10
4. 9, 10, 11
5. 10
6. 4
7. 4
8. 11, 12, 13
9. 10
10. 2

Test 12
1. 15, 17, 18
2. 7
3. 2
4. Approx. 4
5. 4
6.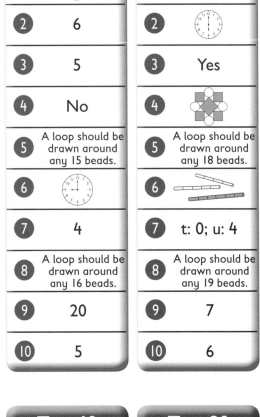
7. No; No
8. 13
9. 3
10. 4

Test 13
1. 17, 18, 19
2. 3
3. No
4. A ring should be drawn around any two cars.
5. A loop should be drawn around any 12 beads.
6. 10
7. 3
8. A loop should be drawn around any 13 beads.
9. Yes
10. 5

Test 14
1. (clock)
2. 6
3. 5
4. No
5. A loop should be drawn around any 15 beads.
6. (clock)
7. 4
8. A loop should be drawn around any 16 beads.
9. 20
10. 5

Test 15
1. 2
2. (clock)
3. Yes
4. (shapes)
5. A loop should be drawn around any 18 beads.
6. (rods)
7. t: 0; u: 4
8. A loop should be drawn around any 19 beads.
9. 7
10. 6

Test 16
1. (clock)
2. 3
3. 12
4. (clock)
5. 5
6. A ring should be drawn around any two boats.
7. 3
8. 14, 15
9. 10
10. 5

Test 17
1. Emma
2. Sally
3. 12
4. 15
5. 10
6. (clock)
7. 3
8. 5
9. Sunita
10. John

Test 18
1. 11
2. 6
3. 16
4. (clock)
5. 6, 8
6. 4
7. (clock)
8. 6
9. Dan
10. Jade

Test 19
1. Triangle
2. (clock)
3. 17
4. t: 4; u: 0
5. 7
6. (boat)
7. (dominoes)
8. 10
9. Emily
10. Seth

Test 20
1. 3
2. 4
3. 18
4. (gate)
5. t: 2; u: 5
6. 20
7. (rods)
8. 10
9. Peter
10. Amaya

Answers

Test 21
1. 5
2. 13
3. 19
4. 11
5. 13
6.
7. 4, 10
8. 10
9. Anil
10. Jane

Test 22
1. 12
2. 8
3. 20
4.
5. t: 2; u: 8
6. Paula
7. 21
8. 10 + 10
9.
10. 17

Test 23
1. 10, 20, 25
2. Yes
3. 21
4. 10
5. 25
6. A ring should be drawn around any three balls.
7. 23
8. No
9. $\frac{1}{2}$ past 8
10. 16

Test 24
1. 28, 34
2. 30, 40
3. 25
4. 10
5. 15
6.
7. 8
8. 30, 35
9. 10
10. 6

Test 25
1. 1
2. 12, 14, 15
3. 10
4. 29
5. t: 2; u: 0
6. t: 2; u: 0
7. 20, 24
8. 6
9. 8
10. 9

Test 26
1.
2. 7
3. A ring should be drawn around any four fish.
4. 1
5. 12
6. 30
7. 31
8. 9
9. 2
10. 1

Test 27
1. 37
2. 7
3. A ring should be drawn around any four sweets.
4. 39
5.
6. 12
7. 40
8. 34, 36, 38
9. 2
10. 6

Test 28
1.
2. 9
3. 35, 40, 45, 50
4.
5. 8; 7
6.
7. 41
8. 9
9.
10. 9

Test 29
1. 46
2.
3.
4. 4
5. $\frac{1}{2}$ past 7
6. A
7. 30
8. 40, 45
9. Any two of the four squares should be shaded.
10. 5; 2

Test 30
1. $\frac{1}{2}$ past 8
2. 10
3.
4. 10; 10
5. 10
6. $\frac{1}{2}$ past 9
7. 49
8. 8
9. 12
10. 3 (accept 4)

47

Answers

Test 31
1. 10; 10
2. 10:30
3. A
4.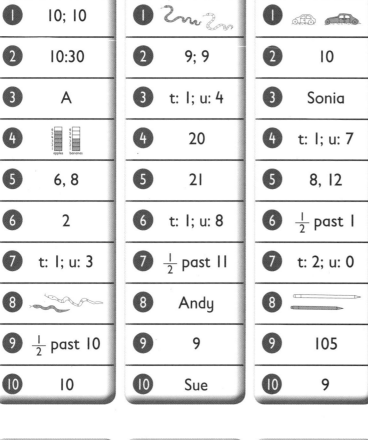
5. 6, 8
6. 2
7. t: 1; u: 3
8.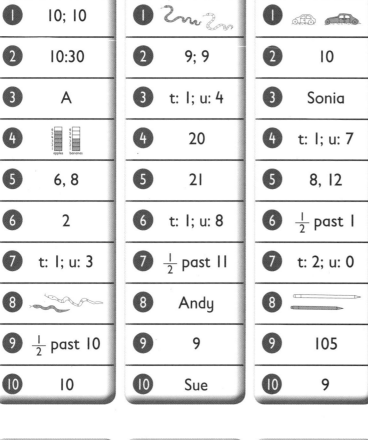
9. $\frac{1}{2}$ past 10
10. 10

Test 32
1.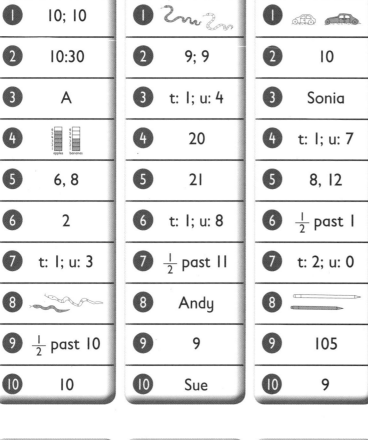
2. 9; 9
3. t: 1; u: 4
4. 20
5. 21
6. t: 1; u: 8
7. $\frac{1}{2}$ past 11
8. Andy
9. 9
10. Sue

Test 33
1.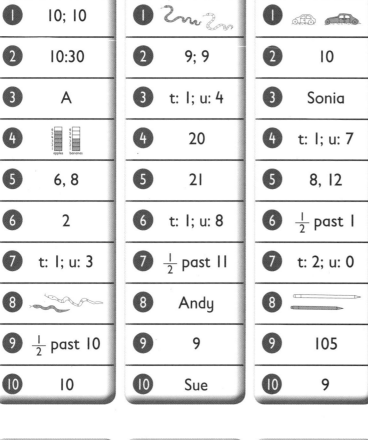
2. 10
3. Sonia
4. t: 1; u: 7
5. 8, 12
6. $\frac{1}{2}$ past 1
7. t: 2; u: 0
8.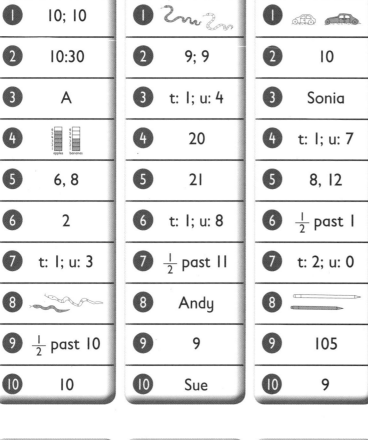
9. 105
10. 9

Test 34
1.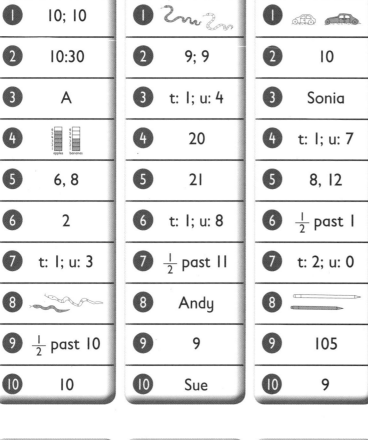
2. 99
3.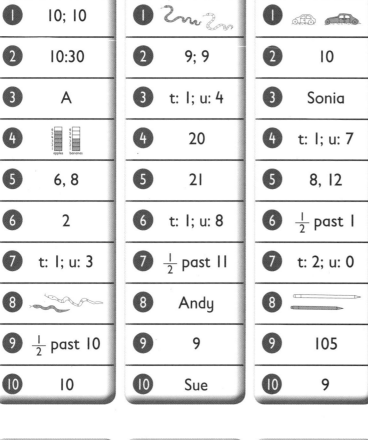
4. 3
5. 1
6. t: 2; u: 1
7.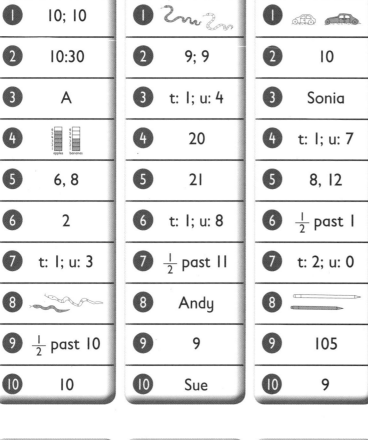
8. 14
9.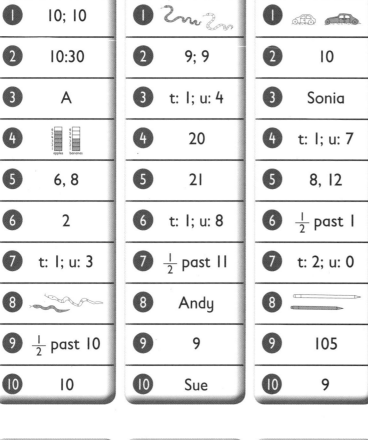
10. 8

Test 35
1. Any two triangles should be shaded.
2. 197
3.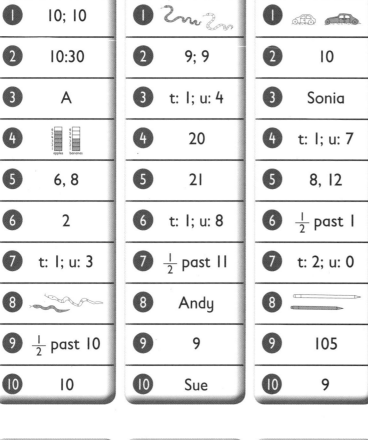
4. t: 2; u: 4
5.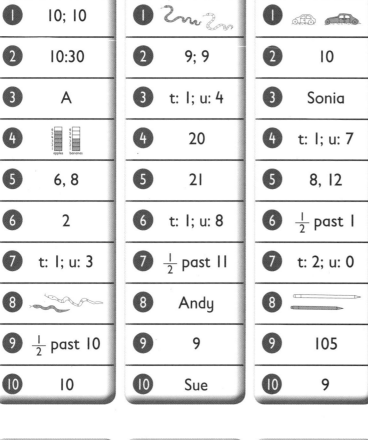
6. Keith
7. (a)
8. 9; 9
9.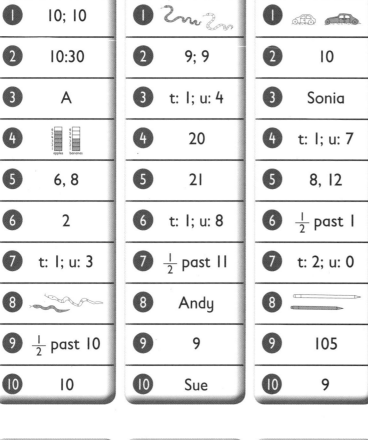
10. t: 2; u: 6

Test 36
1. 8
2.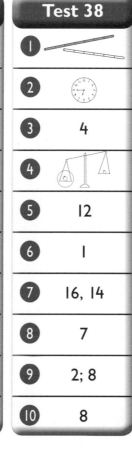
3. 10
4. t: 3; u: 0
5. 300
6. 1:15
7. Approx. 5
8. 31
9.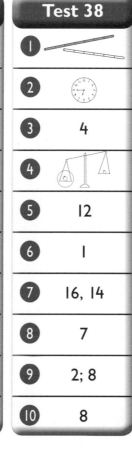
10. 5

Test 37
1. 11
2.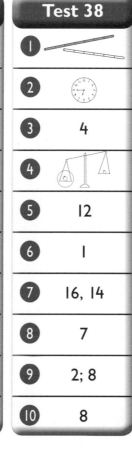
3. t: 4; u: 0
4. Lauren
5.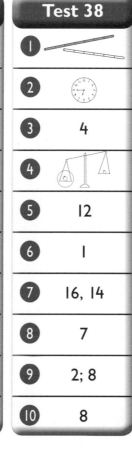
6. 6
7. 5
8. Ross
9. 49
10. 27

Test 38
1.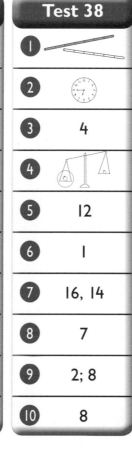
2.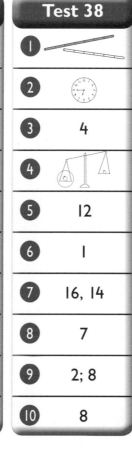
3. 4
4.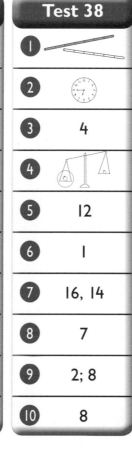
5. 12
6. 1
7. 16, 14
8. 7
9. 2; 8
10. 8

Test 39
1.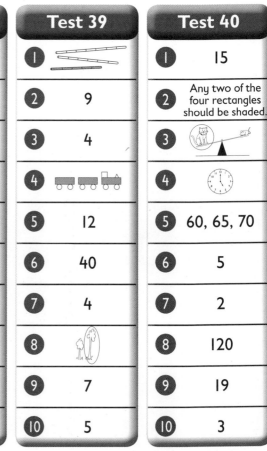
2. 9
3. 4
4.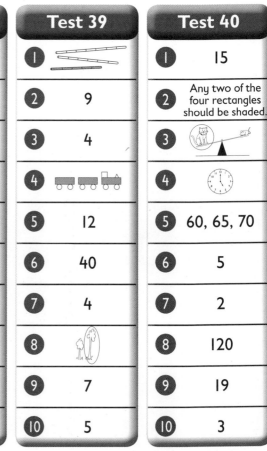
5. 12
6. 40
7. 4
8.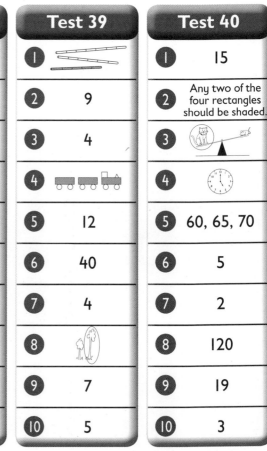
9. 7
10. 5

Test 40
1. 15
2. Any two of the four rectangles should be shaded.
3.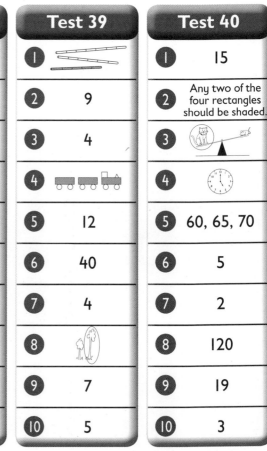
4.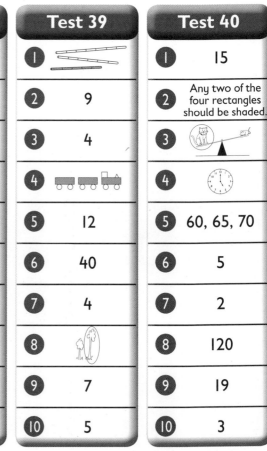
5. 60, 65, 70
6. 5
7. 2
8. 120
9. 19
10. 3

48